Animal Planet

who's for dinner?

predators and prey

wh’s din

DISCOVERY CHANNEL PUBLISHING
Discovery Channel Publishing

Crown Publishers, Inc., New York

for

By Rita Thievon Mullin

Illustrated by Anders Wenngren

ner?

predators and prey

Published by Crown Publishers, Inc., a Random House company,
201 East 50th Street, New York, New York 10022
http://www.randomhouse.com/

CROWN is a trademark of Crown Publishers, Inc.

Printed in the United States of America

Library of Congress Cataloging-in-Publication Data
Mullin, Rita T.
Who's for dinner? : predators and prey / by Rita Thievon Mullin ;
illustrated by Anders Wenngren.
p. cm.
Includes index.
Summary: A collection of facts about the eating and hunting habits of many different animals.
ISBN 0-517-80004-7 (pbk.) — ISBN 0-517-80005-5 (lib. bdg.)
1. Predatory animals—Juvenile literature. 2. Predation (Biology)—Juvenile literature. [1. Predatory animals. 2. Animals—Habits and behavior. 3. Animals—Food habits. 4. Animals—Miscellanea.] I. Wenngren, Anders, ill. II. Title.
QL758.M85 1998
591.5'3—dc21
97-42899

Inspired by Animal Planet, the cable television network from the people who bring you the Discovery Channel. For more information about Animal Planet's availability on your basic channel line-up, contact your cable or satellite company and visit our web site at *www.animal.discovery.com.*

Art Director: David Cullen Whitmore
Research Assistant: Tarah Grant

All photographs supplied by
Bruce Coleman, Inc.

Credits from left to right are separated by semicolons, from top to bottom by dashes.

Cover: Mark Newman—Carl Roessler. Title page: John Visser. 6, 7: Robert L. Fleming. 9: Eckart Pott. 10: Joy Spurr. 13: Erwin & Peggy Bauer. 15: Erwin & Peggy Bauer. 17: Kim Taylor. 18: Carl Roessler. 21: Rod Williams. 22: Rod Borland. 25: Michael Fogden. 27: Gary Meszaros. 28, 29: John Shaw. 31: John Visser. 32, 33: L. Sefton. 34, 35: John Visser. 37: Peter Davey. 38: K & K Ammann. 41: John Shaw. 42, 43: Ron & Valerie Taylor. 44: Jane Burton. 46: Jane Burton—Ed Degginger. 47: Kim Taylor; Joe McDonald. 48, 49: David Madison. 51: Ron Sefton. 53: J & D Bartlett. 54: Tom Brakefield. 57: S. Jonassion. 58: Michael Fogden. 61: Jane Burton. 62: Laura Riley. Back cover: S. Jonassion.

For Michael
and Matthew
— R.T.M.

Contents

who's for dinner?

on the pro

animals stalking prey

wl

Hardly Lion Around
Once a lion has slowly, silently stalked its prey, it makes a deadly leap at a speed faster than 130 miles per hour.

Attack!

Shoot!
A fish called an arrow blenny is only two inches long, but packs a wallop. As it waits in the coral reef for suitable prey, it curls its tail up tightly. When it spots a likely target, it quickly releases the springlike tail against the coral, sending itself flying like a bullet toward its next meal.

Two-Ton Tank
The hippo kills more humans each year than any other African animal—and it's actually a vegetarian! Ornery and likely to charge at the least provocation, the hippo is surprisingly fast on its feet, able to outrun most humans.

Synchronized Swimmers
Pelicans will swim together in a circle and drop their beaks into the water at the same time. These choreographed maneuvers leave an unwary fish with no escape.

Spider Fighter
The tarantula hawk is a wasp that takes on the dreaded hairy spider. A female lures the tarantula out of its home by wiggling the web that covers the entrance. When the spider emerges to retrieve its dinner, the wasp has a brief advantage over the far-sighted spider. The wasp quickly delivers a paralyzing sting, then drags the spider away to a burrow and lays an egg on top of it, providing her larva with the only food it can eat.

◀ **Worth Waiting For**

The kite, a small hawk, loves to eat snails. Instead of breaking open the shell, though, the kite waits patiently until the snail pokes out its head. Then it grabs the morsel in its beak and yanks it out.

Go Fish

Spawning salmon have not only fast-running rivers to fight against but some very hungry bears as well. A brown bear may run into the water, splashing wildly, and chase a fish into shallow water, where it can grab it. At other times a bear may swim underwater to snare a fish swimming past. A scientist keeping count once watched a single brown bear catch 50 salmon in one day.

◀ Star Hunter

The starfish quite literally gets its arms around its dinner. Each arm is covered with suckers that grasp a shellfish. It forces the shell open and positions its mouth over the opening. Then it pokes its stomach inside the shell, releasing a powerful chemical to liquefy the innards. A few sips later, the shellfish is merely a shell.

Sticks and Stones May...

Chimpanzees sometimes throw rocks to kill small monkeys and other prey.

Laughing All the Way to the Bank

The hyena has long been thought to be a scavenger that lived off the kills of other hunters. But scientists following packs of spotted hyenas have learned that they hunt not only small prey but wildebeests and zebras as well. They hunt at night and by morning, the patrol has eaten its fill and may grudgingly share its quarry with a lion or other large predator.

▼ White Out

When sneaking up on prey out on the ice, a polar bear will cover its black nose with its white paw to blend in better with the snow.

Tricks and Traps

Hypno-Ants
To capture slaves from other ant colonies, a species of ant living in Scandinavia sends in a strike force to release a scent that causes the ants to erupt into a fighting frenzy. In the resulting melee the wily invaders release a second scent that lures worker ants to follow them back home.

Bubble Trouble
A humpback whale lassos together a meal of krill, the tiny crustaceans on which it lives, by swimming in a circle beneath a swarm of its prey, blowing bubbles as it goes. The krill are trapped inside the continuous circles of rising bubbles, and the whale sucks them into its mouth.

◀ **Shish Kebug**

Chimpanzees strip sticks of their leaves and drive them into termite mounds and anthills. The angry residents that attack the invading stick are carried out to the chimp's waiting mouth.

Sand Trap
The larva of the ant lion attracts its ant prey by digging a hole in the sand. It uses its head and jaws to flick sand over the side of the hole, then hides at the bottom of its cone-shaped trap. When an unsuspecting ant stumbles over the loose sand, the ant lion uses its head once again, tossing more sand at the helpless ant, causing it to slip to the bottom of the hole and into the larva's ready jaws.

▲ Water Logged
An American alligator will lie in wait in the water, looking for all the world like an old log, until its prey ventures nearby. Then the alligator springs into action, slamming shut its jaws and dragging the struggling victim underwater.

Learner's Permits

Dive-in Theater

Peregrine falcon parents spend several months training their young to dive down and catch prey. One parent will fly over the young bird and drop a meal for it to catch. The other parent flies below to retrieve it if the youngster misses.

Beach Week

An orca parent teaches its young whale pups to beach themselves in order to catch seals in the surf. It will toss a dead seal to a stranded youngster so it can learn to flip itself back into the water with prey in its mouth.

Comfort Food for Sharks

A lemon shark, born in a quiet lagoon in the Caribbean, immediately seeks shelter in the tangle of mangrove tree roots and undergrowth near shore, where it stakes a claim to a narrow region. There it survives for a year or so on small fish and shellfish swimming by. The following year it expands its territory and its list of prey. It moves into deeper water to hunt by day, but returns to its quiet lagoon at night, where it searches for more familiar food.

◀ Tiger See, Tiger Do

Tiger cubs acquire their hunting techniques by following their mother around the Indian countryside. They quickly learn to direct prey toward her and later practice their own kills on prey she has caught. By the age of 18 months they begin to hunt small prey such as rodents. By age three they've learned what they need to know to survive, and they leave their mother to find their own hunting ranges.

▲ **School Uniforms**

After young penguins grow adult feathers, they are ready for school. They spend up to four years in groups learning to fish and swim.

Built to Kill

▼ Anti-Snake Skin
The short-toed eagle is well equipped to protect itself from its favorite prey, a snake in the grass. In order to withstand snake venom, its legs are covered with thick scales and its wings are coated with plentiful down. In fact, before the eagle grabs the snake, it spreads its wings to encourage it to strike where no harm will be done.

Tongue Lashing
The lion's tongue is covered with tiny, hooklike projections. It's so rough that a lion can scrape meat from bones just by licking them.

Dangerous Drivers
Africa's meat-eating driver ants have such a vicious bite that an army of them can strip a cow to the bone in less than a day.

A Bite to Eat
The ten-foot-long requiem shark, which feeds on fish and whale carcasses, bites with a force of 49 tons per square inch.

Call Me Spike
Mantis shrimp, which live in shallow parts of tropical oceans, come equipped with front legs that they fold in front of themselves the way a praying mantis does. But when a shrimp or small fish swims by, those innocent-looking limbs show their true nature: Each is studded with miniature spears that impale the victim, which ends up as a meal in eight-thousandths of a second.

◀ **Dragon-Eye**

The eye of the dragonfly, the most complex of all insect eyes, is made up of 30,000 different lenses, all the better to see potential prey.

Who Needs Dentists?

A shark may replace its teeth every week or two, adding up to more than 30,000 teeth in its lifetime.

Built to Kill

Float Like a Butterfly, Sting Like a Wasp
A wasp uses its sting for more than protection. Although adults live on nectar and other plant products, wasp larvae need protein. To feed its hungry young, the adult attacks other insects, paralyzing its prey with venom from its sting or killing it with a lethal bite before taking it back to the nest.

▼ Sibling Rivalry
If food is scarce, some spadefoot toad tadpoles born in a single batch of eggs will turn cannibalistic. Each "cannipole" grows larger than the average tadpole, and develops a monstrous-looking head with a mouth that looks like an ad for *Jaws*.

A Clean Pate
Vultures have featherless heads and necks, which makes cleanup easier after dining on a carcass.

A Tight Squeeze
An eagle's legs and feet are so powerful that it can crush the bones of its prey with a squeeze of its talons.

One-Way Door
The teeth of the angler fish are attached to its jaw with a hingelike ligament that lets them bend inward as prey is sucked in but spring back up to prevent escape.

Teamwork

▼ Eyes and Ears

The African hornbill and the dwarf mongoose offer prey a one-two punch. The bird stands watch, calling out at danger, while the mongoose heads into the bush, scaring insects and small animals out of hiding. The hornbill snatches any critters that slip past the tiny hunter. If the mongoose oversleeps, the hornbill will give a loud wakeup call down the entryway to its underground home.

The Offensive Line

Chimpanzees hunt in teams, with each member having a different duty. One chimp, the driver, chooses a monkey as prey. Chasers join in the pursuit, and blockers cut off any paths of escape. The monkey is forced toward a waiting ambusher, who captures it.

Leader of the Pack

Wolves live in packs led by a single mating couple. They typically hunt over a 40-square-mile range, which they mark with their urine. When hunting small game, the hunting pack may number only six or seven, but if tracking larger prey, the pack may grow to as large as 20. They can chase at speeds of more than 40 miles per hour and can leap a distance of over 16 feet.

Bring 'em In, Move 'em Up

Sometimes two beaks are better than one. Groups of cormorants and pelicans will often fish together. On the surface, the pelicans herd fish into a circle, while below, the cormorants dive under the fish to ensure that they don't escape.

▲ **Dogging Your Step**

A pack of dholes, wild dogs found in Asia, exhausts its prey by tracking it for hours. Working as a relay team, they regularly exchange positions so that fresh dogs are always in the lead.

A-Hunting We Will Go
The jellyfish-like Portuguese man-of-war is actually a colony of organisms. The floating top is made of cells that give off a gas that keeps it inflated. The tentacles are another species covered with stinging cells. They paralyze fish and bring them up to still other organisms, called feeding polyps, that digest the food.

Teamwork

Hey, Where Are You Heading?
Vultures seeing a fellow scavenger flying at high speed know it must have spotted some prey, and they join in the flight.

Red Light, Green Light
When a group of lionesses spots potential prey, they line up next to each other and begin their approach together. If the prey looks up, they all crouch down and freeze.

Mongooses in Motion
Mongooses, which are about the size of a large housecat, can band together into packs of 50 or so to take on large prey, including small antelopes.

▶ **And in This Corner**

After a battle, an injured hippo will be tended by assistants, like a prizefighter who's returned to his corner. But a hippo's corner team includes birds such as the cattle egret, which removes flies and other insects from the hippo's wounds.

Fatal Attractions

Bug Zappers
Thousands of New Zealand glowworms, or fungus gnat larvae, hang threads from the ceilings of caves. From each thread dangle beads of mucus that reflect the glowworms' light. The glittering cave attracts insects like a porch light in summer, but the visiting bugs get caught on the sticky threads, which the glowworms pull in for supper.

Gone Fishing
The heron catches fish like any fly fisherman—with a lure. It floats insects, feathers, even bits of bread on the surface of the water to attract hungry fish.

Knock, Knock
A species of assassin bug in Costa Rica uses the bodies of its termite victims to attract new prey. It waits near the entrance of a termite nest until a worker on cleanup patrol approaches to remove the body. Then it makes its swift and deadly move.

◀ **Tongue Lashing**
The bright red tongue of the alligator snapping turtle resembles a worm to unsuspecting fish. But when one approaches to take a bite of worm, the turtle snaps its powerful jaws and the would-be predator quickly becomes prey.

Flower Power

The flower mantis looks so much like a blossom that insects come to it in search of nectar.

Spiders' Strategies

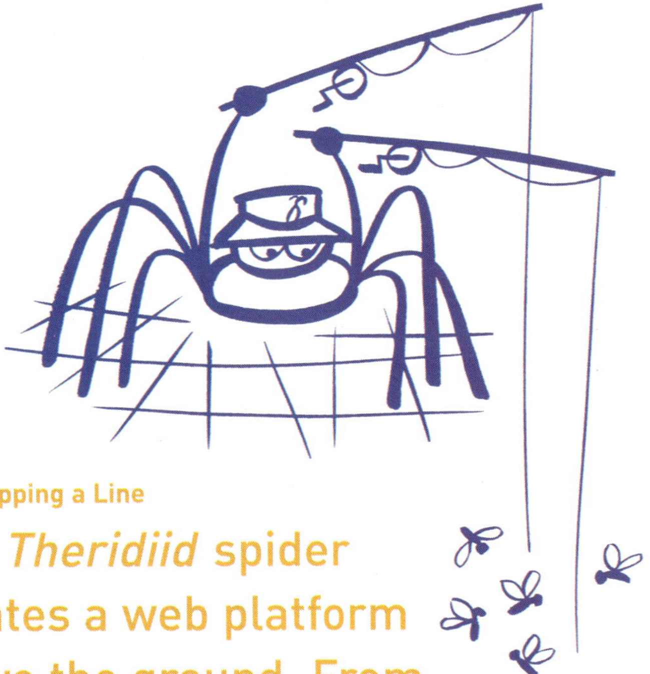

▲ Dropping a Line

The *Theridiid* spider creates a web platform above the ground. From the web it drops down several web fishing lines equipped with sticky bait on the end. When an insect grabs the bait, the line pops up and the spider reels in its dinner.

Surprise!

Rather than setting a trap, the gladiator spider lies in wait for its victims. It hangs upside down, holding a webbed net between its legs. When a potential meal passes by, it drops the net, capturing its victim.

Big, Bad Wolf Spider

The five-inch-long wolf spider found in Costa Rica lives on a diet of frogs. Not only are its fangs venomous, the hairlike spines on its body and legs are tipped with poison that is painful even to humans.

▶ Sewing Machine

The spitting spider doesn't build a web. Instead, when a victim approaches, it spits out a sticky stream as it rocks its head back and forth. Like a thread in a zigzag sewing machine, the line stitches the insect in place so that the spider can approach at its leisure.

▶ **Water Walker**

The raft spider rests on a floating leaf, dipping its front legs into the water to sense any movement from an insect or small fish nearby. It reads the waves of water the way a bat reads pulses of sound in the dark to pinpoint its next victim. Tethered by a silk line, it then walks across the water's surface to nab its prey. The spider paralyzes it with a bite before pulling it onto the leaf to eat. If no prey is near, the spider might even ripple the water itself, attracting curious small fish.

Sensing Food

Hunting Blind

The platypus's soft bill is the control center for a sophisticated tracking system. It is equipped with tiny organs that sense electrical impulses given off by shellfish as they move through the murky water. The system even permits the platypus to hunt successfully at night.

Eye to Eye

A great horned owl, which weighs only about two pounds, has eyes that are the same size as a human's. If human eyes were proportionately as large as an owl's, they would be as big as navel oranges.

◀ **I Hear Dinner Calling**

Woodpeckers can hear a beetle grub chewing wood inside a tree. Once it finds a grub, the bird drills a hole into the wood with its beak and licks out the grub with its long tongue.

▲ **Icy Treat**
A coyote hunting in winter can hear mice moving around beneath the snow.

I'm All Ears
A shark not only depends on its ears to detect sounds, it relies on lines that run along both sides of its body and head that are equipped with receptors for vibrations. These lateral lines permit it to sense sounds made hundreds of yards away.

Smell Like a Fish
Creatures living in dark waters use their sense of smell to find prey. Scientists believe they not only smell with their noses—if they have them—but with special receptors on their skin. Prey try to combat this sense where possible. The inky fluid an octopus gives off when being attacked, for example, hides its smell as well as offering it a cover of darkness.

Bat Snacks

Bug Me
The brown-eared bat eats its own weight in bugs, about 3,000 of them, each night.

▼ **Excess Baggage**
A vampire bat, which is generally only about four inches long, must drink at least half its body weight each day to survive. Its tiny, sharp teeth are so effective that its victims (usually cattle) rarely awaken while the tiny vampire drinks its fill. How does it fly away with the extra weight? Its kidneys get to work immediately, removing the water from the fresh blood. The bat may urinate to get rid of the excess water before taking off.

Just a Click Away
Sonar systems that guide submarines through the oceans owe a lot to bats. Bats navigate through the night sky in search of food without a false move. They produce ultrasonic clicks (too high for humans to hear) and listen for the echoes to detect potential prey—to say nothing of trees, telephone wires, and other obstacles. Different species use one of two methods for flying blind:

Shouters
Many bats produce as many as 200 clicks per second through their mouths. The closer they get to an object, the faster they click, gathering more information about the size and exact location of even the tiniest insect.

Snorters
Other bats snort through their noses, sending longer pulses into the air that are amplified and directed by flaps of skin on their snouts. The bats measure how long the echo of their sounds takes to return to their ears to tell how close they are to an object—and whether it's moving (potential prey) or still (potential obstacle).

I'm All Aflutter

The echolocation system of the horseshoe bat (named for the horseshoe-shaped flap around its nose, which helps direct its sounds) is so sensitive that it can tell the difference between an insect's fluttering wings and leaves blowing in the wind.

defense, de

the art of self-protection

fense

Tougher Puffer
The puffer fish intimidates predators by inflating itself to three times its normal size.

Take Me off the Menu

▼ **Martial Aardvark**

When an aardvark is grabbed by a predator, it does a somersault and kicks its enemy before running to safety.

You Light Up My Life
Most fish that glow live deep in the ocean, but the midshipman is a rare glowing fish that lives in shallow water. Its light, which it uses to scare off predators, is powerful enough to read by.

What a Kick
A giraffe's kick is so strong it can kill an attacking lion.

◀ Gimme Some Skin
The common house gecko sheds a patch of skin when touched, allowing it to escape from attack. It then grows a new layer of skin.

I've Got a Stomach Bug
If attacked, the sea cucumber vomits its insides out. While the curious predator takes a look, the cucumber escapes. It later grows a new digestive tract.

Feeling Wedged In
When a triggerfish is attacked, it heads for the nearest coral reef and slips inside a cranny. Then it raises its trigger, a special backbone, to wedge itself inside. Tug as a predator might, it cannot pull the fish out.

Playing Hard to Get
The klipspringer antelope, which lives in the African mountains, is so surefooted it can leap to a rock no broader than the base of a water glass, giving it a great advantage over big cats that sometimes attack it. Even if a leopard gets close enough to take a bite, it may only get a mouthful of wiry hair. The klipspringer's fur pulls out easily, giving it a chance to escape to rougher terrain.

Take Me off the Menu

Jamming the Frequency
The noctuid moth talks back to bats by sending out its own high-frequency clicks. Like the bright colors on certain frogs and insects that warn predators that they are poisonous, these calls warn bats of the insect's bad taste.

◀ **Sorry, I've Got to Fly**

When attacked, a flying fish can leap out of the water, glide through the air, and land more than 300 feet away.

Blood-Shot Eyes
When a Texas horned toad is threatened, it puffs itself up, which raises its blood pressure so much that the blood vessels in its eyes burst, squirting blood up to seven feet—right at its enemy.

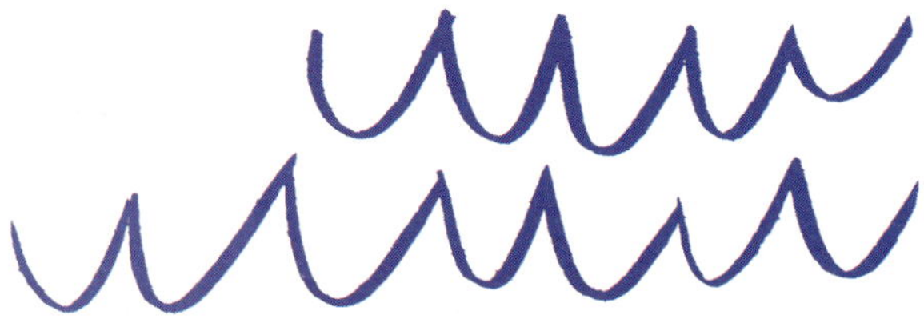

The Bad Housekeeping Award
The ground-nesting hoopoe is one bird that doesn't clean up its youngsters' foul-smelling dung. The odor keeps predators away from its vulnerable nest.

Sailor Fish
A species of sleeper fish in Java floats through the water under a leaf, which it holds with some of its fins while it wiggles others, mermaid fashion, to sail its craft upstream. As it floats along, it is well protected from the kingfisher, a bird that feeds on fish it spots from above. Scientists think the boat may also help the sleeper sneak up on its own prey.

▶ **One if by Land, Two if by Air**
Vervet monkeys (as well as some squirrels and birds) have developed signals to let one another know when a predator is near. The monkeys have different cries for snakes, other ground predators such as big cats, and predatory birds. When they hear the "snake cry," they stand up on their hind legs and look all around on the ground. When the "bird signal" is given, they climb out of the trees and head for the bushes. If a "cat call" sounds, they scramble up the nearest tree.

▲ Howl and Peck

The red-billed oxpecker not only helps out zebras and other large mammals by feasting on the ticks that infest their skin, it also cries out an early warning if any predators approach.

Dr. Fish, DDS

Ocean fish called cleaner wrasses venture fearlessly into the mouths of sharks, serving as aquatic dentists for the fierce predators. There they are safe not only from the shark but also from any other attackers.

May I Help You?

Stop or I'll Sting
The pom-pom crab carries an arsenal of not-so-secret weapons—it wields a prickly sea anemone in each of its pincers to threaten attackers. It holds on to the anemones even when eating. Rather than using its pincers to bring food to its mouth, it uses two of its walking legs.

Porta-Poison
The *Glaucus atlanticus* mollusk is not only immune to the Portuguese man-of-war's notorious stinging tentacles, it actually eats them. When the poisonous cells, called nematocysts, are digested, they provide the mollusk with its own protective sting.

On the Lookout

A water bird called a thick-knee builds its nest near crocodiles. The bird can then set out to find food for her young babies while the mother crocodile keeps an eye on both sets of eggs.

Sensing Danger

▼ **Bifocal Fish**

Some fish that dwell deep in the ocean have eyes with two lenses that focus on different retinas—giving the fish a range of vision of nearly 360 degrees from its two eyes.

Ear-ly Warning System

The kangaroo rat's eardrum is especially large and amplifies sounds 100 times, compared to the human ear, which increases sounds a mere 18 times.

Hairy Legs

Each back leg of a cockroach has more than 200 hairs that are sensitive to the slightest movement of air—which sends it scurrying back into hiding.

▶ **Nose for News**

No one sneaks up on a black rhino. Not only can its ears swivel around to hear sounds clearly from every direction, its nose is very sensitive to new smells, thanks to nasal passages that take up more room in its skull than its brain.

Radar Detectors

Several kinds of moths have ears attuned to the same frequencies that bats use to detect them. When they sense a bat homing in on them, these flying acrobats start looping and diving like planes taking evasive action in a dogfight, confusing the bat.

Looks Can Deceive

▼ **Quick-Change Artists**

The octopus, cuttlefish, and squid can all change color to match their surroundings in less than a second—faster than any other animals.

Heads or Tails?

A bird typically kills a butterfly by pecking its head. But the hairstreak butterfly has outsmarted some birdbrains with a set of threads on its tail that are more noticeable than the antennae on its head. While the bird attacks the wrong end of its supper, the insect flies away in a flash.

▲ See Me? No, See Weed

The leafy seahorse swims through tropical water undisturbed, camouflaged by leaflike extensions growing from its body that make it look like nothing more than a piece of floating seaweed. Some fish are so well fooled that they approach it to hide there from other predators.

Just Call Me Sandy

The flatfish, like many other bottom-dwelling fish, hides beneath the sandy sea bottom to protect itself from predators. It covers itself by smacking the sand quickly, sending grains floating up into the water. It then lies very still as the falling sand covers it almost completely.

▲ Rocky the Crab
Seafloor-dwelling crabs have pebblelike markings that match the rocky surface so well they seem to disappear.

I'm Not *That* Hungry
One kind of caterpillar spins feces into a blanket to hide beneath to repel predators.

Bark, Don't Bite
One species of lanternfly is marked so that it blends in perfectly with the tree bark on which it usually rests. Should a predator approach, it opens its wings to reveal two enormous eyes. To small potential predators, these look like the eyes of a large, hungry enemy.

Ol' Four Eyes

The four-eyed butterfly fish has a false eye near its tail that tricks predators into biting the wrong end. And the butterfly fish's backward-pointing spines make swallowing it tail-first a painful dining experience for an enemy.

Get to the Point

When a predator nears, the spiny anteater digs a quick hole to hide in, leaving only its spines above the ground to pierce a curious predator's snout.

Color Me Poisonous

Many of the most poisonous insects and reptiles (such as yellow jackets and poison-arrow frogs) are brightly colored to warn predators away. One unpleasant experience is all a young predator needs to learn a lesson—if it doesn't know it instinctively. Some non-poisonous animals take the warning a step further, mimicking the markings of poisonous species. Some cockroaches mimic poisonous ladybugs, and the organ-pipe shovel-nosed snake mimics the Arizona coral snake.

Winging It

As its name suggests, the glass-winged butterfly has transparent wings that blend in with any surroundings.

▼ **Stinger Singer**

One cricket that lives in South America not only has markings that resemble a poisonous wasp's, it points one of its back legs out straight to look like a stinger.

Leaf Us Alone

Some insects are so good at blending in that they are practically invisible both to predators and prey. The insects shown in photos 1 and 2 are among the nearly 3,000 different species of stick insect found around the world—although spotting them on a branch can be very tricky. The "thorns" on the branch in photo 3 are really bugs called treehoppers. And the foliage in photo 4 has legs. It's called the wandering leaf.

3

food, glorio

wild dining experiences

us food

What Did You Do Today?
The elephant spends 16 hours a day eating the 300 pounds of grass and fruit it needs to survive.

Animal Pantries

▼ Save It for a Rainy Day

A single gerbil can store more than 100 pounds of food in its burrow.

Ants 1, Grasshoppers 0

North American desert ants don't save for a rainy day, they save for a dry day. They are so good at storing food that colonies have survived through droughts lasting 12 years.

Chubby Cheeks

Chipmunks are well known for carrying food inside their cheeks, but the pocket gopher's cheek pouches open on the outside so it can carry food for later while eating a snack right now.

Bulk Food Aisle

The desert-dwelling kangaroo rat stores plenty of food—a single burrow had 17 cubic feet of seed and dried grass. It also sorts and stores each kind of seed or grass separately.

No Birdbrain

When a marsh tit finds more seeds than it can eat, it hides each leftover seed in a different spot nearby. The bird returns to each location over the next few days to enjoy the hidden treat.

Can You Dig It?

Rather than bringing food to their young, sexton beetles bring their young to the food. When a pair finds a dead animal, they dig a hole under it, lay eggs along the edge, and cover the body with secretions that slow its decay. When the eggs hatch a few days later, the parents eat the food and regurgitate it into their youngsters' mouths.

▲ No Charge for Delivery

It's tough being a hunter when you can't move. Coral depends on currents to carry its diet of plankton its way. But when the sea is quiet, it has a backup food supply ready. Like the leaves on a tree, tiny algae on its surface manufacture food. The algae also supply the coral with its dramatic colors.

Animal Etiquette

▼ Where's the Melted Butter?

The box crab feeds on snails and other shellfish. One of its pincers is shaped like a knife, and the other has a movable thumb for grabbing. When it catches a shellfish, it breaks a hole in the shell with its knife and inserts its other pincer. Then it opens the pincer to crack the shell and stabs the meat inside with one of its legs.

Catch as Catch Can

All oystercatchers eat shellfish, but they don't all have the same table manners. Some hunt underwater for open mussels and stab the meat with their beak. Others use their beak to knock a hole in a shell, then snip the muscle that keeps the shell closed. The method a bird uses depends on what it learned from its parents.

Straw Not Included

The *Lasiodora* giant spider in Brazil is so powerful that it will even attack a small poisonous snake. The spider bites the snake's head, releasing its venom. Then it holds on until the poison has killed the snake. Once its victim is dead, it releases chemicals into the snake's body that dissolve it slowly from head to tail, turning its innards to liquid. Then it sucks the body dry.

Say It with Insects

No self-respecting scorpionfly would consider mating with a suitor unless he presented her with a gift of a large, dead insect.

Time to Eat

Although a single piranha measures only 10 to 12 inches, a large school of these carnivorous fish can kill and consume a 100-pound animal in less than a minute.

It Does a Body Good
The elephant seal pup drinks so much of its mother's milk that it grows at a faster rate than just about any other mammal. A week after birth a baby will have grown by nearly 25 percent.

What an Appetite!

Eating Machine
The shrew, the smallest mammal, must spend most of its time hunting for food. If it stopped eating, it would starve in as little as three or four hours.

Ding Dung
The dung beetle survives on a diet of elephant dung. It also wraps its larvae in dung balls for them to feed on as they grow.

Dining on the Fly
Pipistrelle bats, which capture their insect prey in midair, can consume a quarter of their own weight in bugs in half an hour.

Oh, Baby, What a Meal
A bee larva eats meals of nectar and pollen 1,300 times a day. That's nearly once a minute! By the time it's six days old, it has increased in size 1,500 times.

Eat Like a Bird
To keep on the wing, a bird must eat about 30 percent of its body weight each day—compared with only four percent for a hog or one percent for a hippo.

▶ **Whale of an Appetite**

A 12-ton sperm whale eats about 800 pounds of food each day—the equivalent of 3,200 hamburgers.

Index